Ecological
EVOLUTION OF SPECIES

Do species have ability to conserve their environment?

What leads to perfection in traits of species?

By
Pankaj Vallabh

ISBN-13: 978-1547008148
ISBN-10: 1547008148

Acknowledgements

I sincerely thank my family and friends who have helped me in writing this book.

Preface

It is an incontrovertible fact that organisms have changed or evolved, during the history of our earth. Our understanding of evolution today is based on Darwin's theory and variants of the theory 'Organisms with traits that give them an advantage over their competitors are more likely to pass on their traits to the next generation than those with traits that do not confer an advantage'. However all evolution cannot be explained by Darwin's theory.

Evolution of any species does not occur in isolation but always occurs in an environment and is influenced by that environment. The species need the survival of that environment for their own existence. We will see that an important factor in the evolution process is that the evolving organism do not damage or destroy the environment they live in. By considering this factor, we shall see that the evolution process leads to development of almost perfect traits in animals and organisms, suitable for their role in the environment. Also it will lead to wide diversity in the species inhabiting the environment.

This book discusses the principles which determine, that after mutations or changes, which organisms in an environment survive to carry on their genes to their next generation and thus the underlying principle for the evolution of various species on earth. It will be seen that for evolution, besides fitness to compete and survive; equally important is the ability to conserve the resources it needs for its survival. Species will not be able to survive over a period of time if they damage or destroy the resources they require. Besides survivability, it also addresses sustainability. The major difference with this approach to the development and emergence of organisms as compared to theory of evolution by natural selection is that whereas in natural selection, species and organisms are believed to have evolved by continuous selection from groups of organisms, the better of which are fit to survive; in this approach, the evolving organisms are those which are both fit to survive and let their environment survive. Neither will any species exist which does not have ability to compete and survive along with others in its environment, nor will any species survive over a period of time which cannot conserve its environment. The role of sustainability in survival and evolution of species is as crucial as fitness. Inability to conserve one's environment may be due to the species becoming strong, and growing at the same time, so that it consumes or destroys all of its food sources and habitat, it produces products which are harmful to the growth of its environment, etc..

The continuous addition of new features and changes leads to the emergence of complex and developed organisms. Some organisms may be more complex and more developed than others, but they are all equal from the point of view of their capability to survive in their environment. From the purpose of surviving and growing in an environment, an organism living in an ecosystem or sub ecosystem is equal to all other organisms sharing its ecosystem.

The theory of evolution by natural selection states that organisms evolve by competition and natural selection, where the better organisms are selected from the prevalent group of organisms. There has been evidence that some of the species of fish have been living for millions of years. Why have they not evolved into better and superior creatures? Are they the perfect creatures? With so much variation in traits or diversity within every species, if evolution took place by natural selection i.e. every generation resulted in favorable trait or phenotype being selected, we would see evolution take place at a very rapid rate and not a small change to occur in thousands or even hundreds of thousands of years, as we see it. Similarly, there are many questions, in theory of evolution by natural selection, to which there are no appropriate answers.

The theory presented here owes a great debt to observations and originals insights of Motoo Kimura (Neutral Theory of Molecular Evolution 1985), Amotz and Avishag Zahavi (Handicap Principle 1997) and Stephen P. Hubbell (Unified Neutral Theory of Biodiversity and Biogeography 2001).

This book is written in continuation with my previous book 'Conserve to Survive', the main ideas and chapters of which, have been incorporated in this book.

Contents

Theories of Evolution

Evolution is change over time through descent with modification. Theories concerned with origin of the earth, the universe, and life, are indeed diverse and uncertain. Science, contrary to popular belief, cannot contradict the role of divine being or force in the creation or evolution of early universe or early life[28-31]; no one for certain knows the ways of the divine. Like a seed, slowly and gradually growing and evolving, in an ordered and sequential manner, into a huge tree, it may well be that after the emergence of the first seed of life on earth, it was just a slow and gradual process, in an ordered and sequential manner, taking millions or billions of years, ultimately leading to the emergence of humans.

By evolution we mean the development of life in geological time. It implies an overall gradual development which is both ordered and sequential. In terms of living organisms it may be defined as the development of differentiated organisms from pre-existing, less differentiated organisms over the course of time[28-31].

The major theories accounting for the origin of life on earth[28-31]:

Much of the evidence on which these theories are based is metaphysical, that is to say it is impossible to repeat the exact events of the origin of life in any demonstrable way. This is true from both scientific and religious (theological) accounts. The major theories are:

Special creation: Life was created by a supernatural being at a particular time. This theory is supported by most of the world's major religions and civilizations and attributes the origin of life to a supernatural event at a particular time in the past. Archbishop Ussher of Armagh calculated in 1650 A.D. that God created the world in

October 4004 B.C. beginning on October 1st and finishing with man at 9:00 a.m. on the morning of October 23rd. Some believe that the world and all species were created in six days of 24 hours' duration. They reject any other possible views and rely absolutely on inspiration, meditation and divine revelation.

Ussher

While science broadly relies on observation and experiment to seek truth, theology draws its insight from divine revelation and faith. Faith accepts things for which there is no evidence in the scientific sense. This means that logically there can be no intellectual conflict between scientific and theological accounts of creation, since they are mutually exclusive alms of thought. Scientific truth to the scientist is tentative, but theological truth to the believer is absolute.

Since the process of special creation occurred only once and therefore cannot be observed, this is sufficient to put the concept of special creation outside the framework of scientific investigation. Science concerns itself only with the observable phenomena and as such will never be able to prove or disprove special creation.

Aristotle

Spontaneous generation: This theory was prevalent in ancient Chinese, Babylonian and Egyptian thought as an alternative to special creation. Aristotle (384-322BC), often hailed as the founder of biology, believed that life arose spontaneously. Aristotle's hypothesis of spontaneous generation assumed that certain 'particles' of matter contained an 'active principle' which could produce a living organism when the conditions were suitable. He was correct in assuming that the active principle was present in a fertilized egg, but incorrectly extended this to the belief that sunlight, mud and decaying meat also had the active principle.

Van Helmont

"Such are the facts, everything comes into being not only from the mating of animals but from the decay of earth …And among the plants the matter proceeds in the same way, some develop from seed, others as it were, by spontaneous generation through natural forces; they arise from decaying earth or from certain parts of plants." (Aristotle). Van Helmont (1579-1644), a much acclaimed and successful scientist, described an experiment which gave rise to mice in weeks. The raw material for the experiment were a dirty shirt, a dark cupboard and a handful of wheat grains. The active principle in this process was thought to be human sweat. In 1688, Francesco Redi, an Italian biologist and physician living in Florence, took a more rigorous approach to the problem of origin of life and questioned the theory of spontaneous generation. Redi observed that the little white worms seen on decaying flesh were fly larvae. By a series of experiments he produced evidence to support the idea that life can arise only from pre-existing life, the concept of biogenesis.

Francesco Redi

Steady-state theory: This theory asserts that the Earth had no origin, has always been able to support life, has changed remarkably little, if at all, and that species had no origin.

Estimates of the age of the Earth have varied greatly from the 4004BC calculation of Archbishop Ussher to the present day values of 5000 million years based on radioactive decay rates. This theory proposes that Earth has always existed, and that species, too, never originated, they have always existed and in the history of a species the only alternatives are for its numbers to vary, or for it to become extinct.

The theory does not accept the palaeontological evidence that presence or absence of a fossil indicates the origin or extinction of the species represented and quotes, as an example, the case of coelacanth, Latimeria. Fossil evidence indicated that the coelacanths died out at the end of the Cretaceous period, 70 million years ago. The discovery of

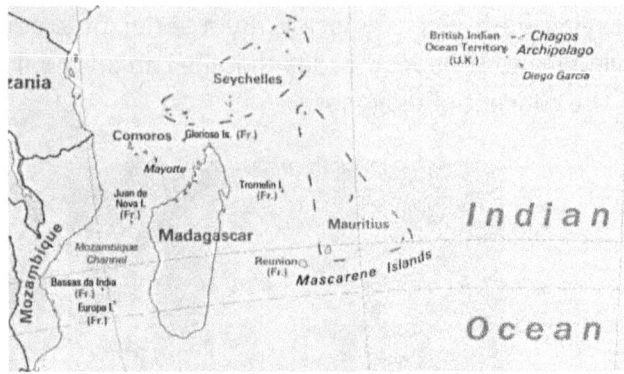

Coast of Madagascar

living specimens off the coast of Madagascar has altered this view. The steady state theory claims that it is only by studying living species and comparing them with the fossil record that extinction can be assumed and then there is a high probability that this may be incorrect. The palaeontological evidence presented in support of the steady state theory describes the fossil's appearance in ecological terms. For example, the sudden appearance of a fossil in a particular stratum would be associated with an increase in population size or movement of the organism into an area which favored fossilization.

Cosmozoan theory: This theory favors the idea that the origin of life could have an extraterrestrial origin. It does not, therefore, constitute a theory of origin, as such, but merely shifts the problem to somewhere else in the universe.

This theory proposes that life could have arisen once or several times in various parts of our galaxy or the universe. Its alternative name is the theory of panspermia. Repeated sightings of UFOs, cave drawings of rocket like objects and 'spacemen' and reports of encounters with aliens provide the background evidence for this theory. Research into materials from meteorites and comets has revealed the presence of many organic molecules, such as cynogen and hydrocyanic acid, which may have acted as 'seeds' falling on a barren Earth. In all cases further evidence is needed.

Biochemical evolution: This theory is based on the emergence of life by complex chemical reactions taking place in the Earth millions of years ago.

It is generally believed that the Earth is some 4.5 to 5.0 billion years old. Many biologists believe that the original state of the Earth bore little resemblance to its present day form and had the following probable appearance. It was hot (about 4000-8000C) and as it cooled carbon and the less volatile metals condensed and formed the Earth's core; the surface was probably barren and rugged as volcanic activity, continuous earth movements and contraction on cooling, folded and fractured the surface.

Alexander Oparin

In 1923 Alexander Oparin argued that organic compounds, probably hydrocarbons, could have formed in the oceans from more simple compounds. The energy for these synthesis reactions was probably supplied by the strong solar radiation (mainly ultraviolet) which surrounded the earth before the formation of the ozone layer. Oparin argued that if one considered the multitude of the simple molecules present in the oceans, the surface area of the earth, the energy available and the time scale, it was conceivable that oceans would gradually accumulate organic molecules to produce a 'primeval soup', in which life could have arisen.

In 1953 Stanley Miller, in a series of experiments, simulated the proposed conditions on the primitive Earth. In his experimental high energy chamber, he successfully synthesized many substances of considerable biological importance, including amino acids, adenine and simple sugars such as ribose. More recently Orgel at the Salk Institute has succeeded in synthesizing nucleotides six units long (a simple nucleic acid molecule) in a similar experiment. Oparin's theory has been widely accepted, but major questions remain in explaining the transition of complex organic molecules to living organisms.

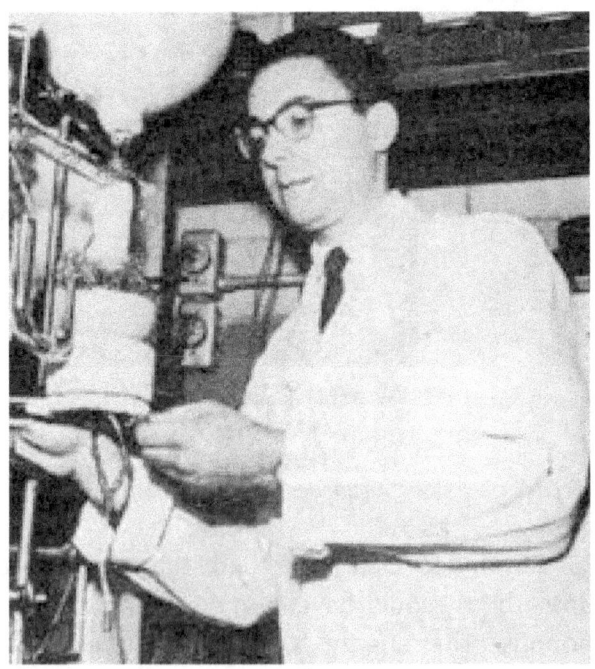

Miller

Evidence of evolution of species[1]: Species is defined as 'groups of actually or potentially interbreeding natural populations, which are reproductively isolated from other such groups'. Evidence of relevance to the theory of evolution is provided from many sources, the main ones being:

Palaeontology: It is the study of fossils. Fossil evidence alone is not sufficient to prove that evolution had occurred, but it supports a theory of progressive increase in complexity of organisms.

Geographical distribution: As early European naturalists explored the newly discovered lands of Africa, Asia and America, they found that the number of species, or different types of organisms, was much greater than anyone had suspected. Naturalists also observed that some of these species closely resembled each other, yet also differed in some characteristics. These observations led some naturalists to consider that species could change after all, and some of the similar species could have developed from a common ancestor.

Classification: Systematists value the similarities that arise when two kinds of organisms share a feature because both inherited it from a common ancestor. Although the sea squirt, a type of tunicate, spends its adult life permanently attached to rocks on the ocean floor, its free-swimming larva has a nerve cord, tail and gills, placing it in the same phylum as the vertebrates phylum Chordata.

Plant and animal breeding (artificial selection): Farmers and animal breeders, both today and earlier, take advantage of the natural variation within a population to select for characteristics which they find valuable or useful. By choosing organisms that naturally exhibit a particular trait and then breeding that organism with another of the same species exhibiting the same trait, breeders are able to produce animals or plants having a desired inherited trait.

Comparative anatomy: Studies have shown that many structural features of different organisms are basically similar. For example, the bones of all mammalian forelimbs are similar in spite of differences in shape.

Adaptive radiation: A species gives rise to many new species in a relatively short time. It occurs when populations of a single species invade a variety of new habitats and evolve in response to the differing environmental pressures in those habitats.

Comparative embryology: Related organisms go through similar stages in their embryonic development. For example, the early stages in limb formation, the limb buds, are virtually indistinguishable in vertebrate embryos. The resemblances lie in more than the external appearance. The arrangement of the arteries and the structure of the developing hearts in early vertebrate embryos also follow a very similar pattern.

Comparative biochemistry: All living things have DNA as their genetic material, with a genetic code, which is almost universal.

Theories of Evolution[28-31]:

Lamarckian Evolution: The French biologist Lamarck proposed, in 1809, a hypothesis to account for the mechanism of evolution based on two conditions: the use and disuse of parts, and the inheritance of acquired characteristics. Changes in the environment may lead to changed patterns of behavior which can necessitate new or increased use (or disuse) of certain organs or structures. Extensive use would lead to increased size and/or efficiency whilst disuse would lead to degeneracy and atrophy. These traits acquired during the lifetime of the individual were believed to be heritable and thus transmitted to offspring.

Lamarck

Darwin, Wallace and the origin of species by natural selection: In July 1858, Darwin and Wallace presented papers about their ideas on origin of species at a meeting of the Linnean Society in London, and in Nov. 1859, Darwin published *On the origin of species by means of natural selection.* Under intense competition of numbers in a population, any variation which favored survival in a particular environment would increase that individual's ability to reproduce and leave fertile offspring. Less favorable variations would be at a disadvantage and organisms possessing them would therefore have their chances of successful reproduction decreased. Darwin wrote: 'As many more individuals of each species are born than can possibly survive, and as such, consequently, there is a frequently recurring struggle for existence, it follows that any being, if it

Charles Darwin

Alfred Wallace
www.wku.edu/~smithch/index1.htm

vary however slightly in any manner profitable to itself, under the complex and sometimes varying conditions of life, will have a better chance of surviving and thus be naturally selected. From the strong principle of inheritance, any selected variety will tend to propagate its new and modified form'. This theory is presently the most widely accepted theory of evolution and is popularly known as *theory of evolution by natural selection* (also *struggle for existence* or *survival of the fittest*).

Chapter 1
Introduction

Evolution is change over time through descent with modification. It is an incontrovertible fact that organisms have changed or evolved, during the history of our earth. Our understanding of evolution today is based on Darwin's theory and variants of the theory 'Organisms with traits that give them an advantage over their competitors are more likely to pass on their traits to the next generation than those with traits that do not confer an advantage'. However all evolution cannot be explained by Darwin's theory.

Evolution of any species does not occur in isolation but always occurs in an environment and is influenced by that environment. The species need the survival of that environment for their own existence. We will see that an important factor in the evolution process is that the evolving organism do not damage or destroy the environment they live in. By considering this factor, we shall see that the evolution process leads to development of almost perfect traits in animals and organisms, suitable for their role in the environment. Also it will lead to wide diversity in the species inhabiting the environment.

Many times we see deadly viruses like Spanish Flu evolve and then die off within a limited span of time. Why are they not able to survive in spite of having better fitness than their competitors? Evolution has been studied by seeing the survivability of organisms under different conditions. What about sustainability? The same organism will not be able to live for long if it cannot sustain itself. What is the role of sustainability in evolution? For explaining evolution, Natural selection deals with survivability, whereas this book addresses both survivability and sustainability.

The theory presented here owes a great debt to observations and originals insights of Motoo Kimura (Neutral Theory of Molecular Evolution 1985), Amotz and Avishag Zahavi (Handicap Principle 1997) and Stephen P. Hubbell (Unified Neutral Theory of Biodiversity and Biogeography 2001).

Prof Zahavi had very correctly observed the handicap or vulnerable traits in many species and interpreted the observation as a sign of greater strength in those members. Zahavi's Handicap principle may be summed up as 'Strength is demonstrated by showing traits of vulnerability'. This is peculiar and quite contrary to evolution theory of natural selection, according to which we would expect to see a member with more strength but with a trait of vulnerability, see the vulnerable trait reduce with time and just the member with more strength but without the vulnerable trait become more prevalent in the species. Even if the member with more strength is not distinguishable from other members, we would still see the member with more strength have a better survival chance and thus with time become more prevalent in the species, as can be expected from natural selection theory. Another way to interpret Prof Zahavi's observation is these handicaps would be a necessary requirement for sustainable living of a species, as is explained in more detail in chapters 6. Dr. Stephen P. Hubbell has defined *Neutrality as per capita ecological equivalence among all individuals of every species in a given trophically defined community. By neutral it means that the organisms in the community are essentially identical in their per capita probabilities of giving birth, dying, migrating, and speciating*[23]. This book takes it further to explain why neutrality is important.

In a long term study of a deer population in a predator free fenced area in Texas found that the deer density increased for a few years and then sharply declined to match the populations outside the predator enclosure[26] and the deer in the predator free area had a much poorer diet as they had overgrazed[27]. Probably this would not have been the consequence, had it been a slow process and deer had time to adapt to the new reality of no wolves or coyotes – they might have developed traits to limit their population, limit their grazing ability, or a handicap, so that an appropriate balance could be maintained between the population of deer and amount of vegetative grass available for grazing.

We see this aspect in nature – those organisms and animals which are least harmful to their hosts are most populace and the deadlier they are, proportionally lower is their population size. Annually in US there are one billion cases of common cold – a mild infection, tens of millions of cases of influenza – a strong infection, and about 17,500 cases of bacterial meningitis – a deadly infection. This is further discussed in chapter 3. While the approach differs from the view, that any change that is beneficial to the organism gets selected, here we take a holistic approach to the process of evolution. Here we take into account that an organism is a part of its ecosystem, and any change which is beneficial to the organism while conserving its ecosystem, gets selected. Using this approach, we will see that **species evolve by becoming more efficient in what they do, leading towards perfection in their various traits.**

The ecological model presented here is a model of evolution where the fitness level is bound within limits. A model of evolution based on perpetual growth and development would be a model where organisms ultimately destroy the very place they inhabit. Sustained growth in fitness is unsustainable. A model of evolution based on natural selection or survival of the fittest would be a similar model of perpetual growth and development and which would ultimately lead to the destruction of the very place they grow. However, this is not so and we see relatively stable ecology over long periods of time. **Natural selection deals with survivability whereas ecological selection, where fitness level remains within limits, deals with both survivability and sustainability.**

In nature, it is quite common to see pair or group of natural enemies like polar bear and seal; snake and frog; wolf, coyote and deer, etc.. While their population have varied overtime, it is rare to see either the prey or predator go extinct. In fact, although the prey and predator have coexisted over thousands, hundreds of thousands or even millions of years neither could obliterate the other. The only reason either of them could not develop to a level so as to obliterate the other, is need of the other for their own existence. However, each has evolved into a better form over time, as can be seen from historical records.

All species are identical on a per capita basis in their probabilities of birth, death, and dispersal[23]. There are neither good nor evil living organisms in the world, but only co-inhabitants, each of which plays an assigned role, and each of which need the survival of others for their own survival. The living organisms remain in a limited band of capability, below which they are not able to survive, and above which they become detrimental to the resources in the environment they need for their survival. So remaining within its limited band of capability, the only way to evolve and increase its population is to increase the efficiency in each trait, thereby doing the same activity with lesser requirement of resources – thereby bringing about perfection in its different traits. This is exactly what we see in nature – organisms and animals with almost perfect development of their features best suited for their requirement.

An implication from this book can be that species have the capability to evolve into the most deadly form or most docile form depending upon its environment. So modern medicines, antibiotics etc. just push organisms to evolve into their more deadly forms requiring perpetual development of new medicines.

Some unanswered questions from *The Theory of evolution by natural selection:*

In the theory of evolution by natural selection, there are many questions to which there are no appropriate answers.

If the new emerging organisms were better and were continuously selected from the previous group of organisms, it would lead to the extinction of the group of organisms from which it emerges, thereby destroying its former simpler forms and leading to the emergence of a single superior and more developed organism with all simpler and lesser developed organisms becoming extinct. However, this is not true, and in our environment, we see the simple organisms surviving and coexisting along with more complex and developed organisms.

If by natural selection, the best are selected and propagate, then why is there diversity among individuals of a species, i.e. why not only the best within the species are selected to survive. In fact, even in a species, we will rarely find any two individuals of a species identical.

People ask, if organisms evolve into better and superior species and individuals than the previous species and individuals, and if humans evolved from monkeys, then why do we still see the monkeys. Also why did the intermediate or early forms of humans disappear whereas we still see the monkeys and gorillas?

If by evolution, superior beings are evolved, then why do humans get killed by a bite from a cobra, get overpowered by wolves, get crushed by a python, cannot confront a lion, or get killed by anthrax microorganism?

If big changes in organisms like new features are achieved by very slow and gradual accumulation, how does an individual in a species with a minute change stand out from others in the species and go towards increasing the new change, if the change itself is very minute and will contribute very little towards its survival.

With a new and useful feature, the selection of an individual from a group, would be more justified if all the individuals of the group were identical i.e. there is no diversity in the group. But there is diversity in all species and for any individual in the group there will be some in the group with better features and some with lesser features than itself. Thus if an individual gets a small new useful feature, how will it stand out from others in the group and go towards increasing the new change if there are others in the group which have better features than itself?

Millions of years ago, when life started to appear on earth and various life forms began to occupy the different areas of the environment, if the organisms became better by competing for the same resources, then how were they able to expand and occupy the different ecological areas. For example how did an organism living on the ground evolve into a bird and start flying if it only competed with fellow organisms for resources on the ground?

Fossil of a 40 million year old fish

There has been evidence that some of the species of fish have been living for millions of years. Why have they not evolved into better and superior creatures? Are they the perfect creatures?

With so much variation in traits or diversity within every species, if evolution took place by natural selection i.e. every generation resulted in favorable trait or phenotype being selected, we would see evolution take place at a very rapid rate and not a small change to occur in thousands or even hundreds of thousands of years, as we see it. However, when there is a change in equilibrium state, like when an organism moves or is moved from its original environment to a new environment, we do see a rapid rate of evolution and rapid changes in the organism.

In the theory of evolution by natural selection, saying that organisms with better genes are selected, could be justified if all organism are of the same age (or in similar age as compared to their life span i.e. all organisms are fully matured, or all organisms are of 50% maturity size or all organisms are of 60% over the maturity age). However in reality the organisms which normally get eliminated are the young, which are the underdeveloped, or the aged, which are the weak.

Scope of this book: This book does not discuss how life first emerged or how changes take place in living organisms due to mutation, adaptation, etc., but discusses the underlying principle for the survival and growth of organisms in its environment. Thus after the first emergence of life and changes taking place in the organisms due to mutation, adaptation, etc., it discusses the underlying principle that determines which organisms survive to carry on their genes to their next generation; thus it discusses the underlying principle for the evolution of various species on earth.

Chapter 2
Why the common cold is so common

Is it not surprising that the common cold virus is quite prevalent; whereas, influenza – a stronger virus, is less prevalent; and still a more strong virus – Spanish flu, just cannot survive and fades away in a short span of time. This would not support the theory of natural selection as we see a stronger and thus better fit organism not able to survive whereas its milder form survives and flourishes.

The common cold is so common, probably because the common cold virus can stay in equilibrium with its environment, like in the human body. The virus while staying and growing in the human body, does not destroy its environment i.e. the human body, its source of shelter and nourishment; It flourishes and jumps to other humans, surviving and growing, while at the same time keeping its environment intact. On the contrary, deadly virus or bacteria, after infecting a body, if they destroy it, they destroy their environment and source of nourishment. By destroying all bodies they infect, they destroy their environment and all sources of their nourishment, and ultimately destroy themselves. However, in due course of time, their hosts from other parts of the environment would migrate and again inhabit the area, returning it to its original state, but without the deadly organism. For example, in viruses infecting rabbits, if a virus evolved which would kill all rabbits; it would result in the elimination of all rabbits in the area; with no rabbits, it would not have any source of shelter or nourishment, resulting in the elimination of the deadly virus itself. However, in due course of time, rabbits from other areas would begin to inhabit the area, and thus the area would return to its original state but without the deadly virus. Only those deadly organisms will survive which can stay dormant for long periods till their environments revive, or those which jump to other bodies not very quickly, i.e. infect only a few in the population.

Rabbit

Thus it is not only important to be fit to survive in the environment but also ensure that they do not damage their environment. This will be true for almost all organisms.

Are we superior?: More complex? Yes; More developed? Yes; But superior? I would not say so. Otherwise how can we explain death by deadly diseases like anthrax caused by micro organisms. The micro organisms are small, simple in construction, probably one of the earliest forms, yet powerful enough to kill humans. Although humans are superior in intellect, many other organisms are superior to humans in others ways, e.g. speed, strength, adaptability to adverse conditions, adaptability to different environments (like birds in the air, fishes in the water, etc.).

Love, fear, habits, customs and social behavior all contribute to the total features of an organism that help it to survive in the environment. For example it would not be possible for a human baby to survive if it did not have the love and care of its parents, or for humans to survive if it did not have fear and reach for safety on sight of a lion, or for a group of wolves to survive if they did not form coordinated groups while hunting or defending themselves. Birds fly south in winter in well organized formations which help them fly long distances, penguins carry their eggs on their feet till they hatch, which protects them from the cold, etc.. The relative fitness of all species in an environment must be the same. *For fitness invariance there are different trade-off combinations of life-history traits that confer equivalent per capita relative fitnesses on all the species exhibiting them. This must be true by an almost self-evident proof. All species that manage to persist in a community for long periods with other species must exhibit net long-term population growth rates of nearly zero[23].*

Black crowned night heron bird

Photo: Jessie Cohen © http://nationalzoo.si.edu

Thousands of years ago, when in Egypt, we had highly developed society of pharaohs; during the same period, a few thousand kms south, we had very primitive societies living in jungles with no developed language. Isn't it surprising that while this highly developed society collapsed, the primitive societies of south and middle Africa still exist today. My point here is that to be advanced, have better capabilities or more intelligence is not the only criteria for evolution.

Darwin's theory of evolution by natural selection can be viewed as evolution taking place due to improvement in capability of organisms, whereas the approach in this book is that evolution takes place by improvement in features of organisms accompanied by features which ensure that they do not destroy or damage the available resources on which they depend. Greater capability achieved by the improved features can lead to organisms damaging or destroying their environment. With the help of examples, it is shown that capability to conserve or retain the required resources of the environment is essential for the process of evolution or struggle for survival in any organism. Species have to be ecologically fit.

Chapter 3
Conservation for survival

Various articles, papers and books have been written on the importance of conservation of environment, ways to conserve environment, and the effect of change in environment on the evolution of new diseases, extinction of some species, etc.. But here we examine the feature in an organism to conserve its environment and the role of this feature as a factor in its survival and in the evolution of the species. It is shown, using population data, historical evolution, extinction of organisms, and characteristics of species, that the 'feature in a species to conserve its environment' would play a very important role in the survival of the species and in the evolution of new species.

An important factor in the survival of any existing species and evolution of a new species is the feature in the species to conserve its food source and the environment in which it grows, directly or indirectly. Those that do not have the capability to conserve their environment fade away and are ultimately eliminated. Thus many evolving organisms, in spite of being stronger and better than others in the environment, do not form a part of the evolution process. Changes take place in organisms due to mutation, genetic drift, adaptation, migration, etc. and natural selection has been considered to be the major factor in evolution. However, evolution of all species in nature cannot be fully explained by natural selection and the feature to conserve their environment could be an important factor in the evolution of a species.

Although everyone agrees with the need and importance to conserve their environment, probably very few take it seriously. We see countries and multi-national companies, while advocating conservation expect others to do so while at the same time strive for their own growth which ultimately comes at the cost of the environment. While everyone agrees with the need to preserve our environment, we see annual increases in petroleum production, human population, deforestation to give way to farming and residential areas, production of pesticides, insecticides, plastics, other indestructible pollutants, etc.. This chapter further stresses the need to conserve our environment. So while exploitation of resources available by nature is important, ultimately over a long period of time, probably equally important or even more important is the need to conserve our environment.

If we examine the environment, we will find that all organisms have features to conserve their environment – whether, by being vulnerable to predators but at the same time having a high growth rate, producing excreta and products which are well utilized by others in their environment, those strong and not vulnerable having a limited growth rate, etc.. *In a comprehensive 15-year study of white-tailed deer and wolves, the Minnesota Department of Natural Resources monitored the movement, survival and mortality causes of deer and wolves. Wolves end up surviving primarily on the most vulnerable individuals in the deer population, such as the very young, old, sick, or nutritionally compromised deer, because those are the ones they catch. Despite the fact that deer outnumber wolves by 150 to 1, wolves are not particularly effective hunters of white-tails[22]*. One way of interpreting this study is that wolves are not effective hunters and have scope to evolve and become better hunters; another way of interpreting the same study is that the capability of wolves is sufficient for them to survive and also to maintain a healthy supply of deer for them in their environment. If they had become better hunters, probably they would not be able to maintain sufficient supply of deer for them to survive. So the only way for them to evolve further would be to have the same capability with a lesser requirement of food i.e. increase in efficiency in its various traits. Organisms have feature to conserve their environment – those that do not have this feature cannot sustain themselves and ultimately become extinct.

When an organism develops a new useful feature, it is important that it has features to conserve its environment - by a decrease in its other features, so that it is equal to others in its environment and does not damage others; by developing features so that its growth remains controlled and limited; producing products which can be consumed by its environment – otherwise in due course of time its product will fill the environment and damage the environment in which it grows; etc.; Like many deadly diseases e.g. similar to plague or deadlier, which frequently arise and fade away within a short period of time, often called dying a natural death, these deadly diseases in spite of having features which make them better and stronger than others in their environment, do not form a part of the evolution process. *As per World Health Organization, three times in the last century, the influenza A viruses have undergone major genetic changes mainly in their H-component, resulting in global pandemics and large tolls in terms of both disease and deaths. Spanish flu outbreak in 1918-1919 killed up to 40 million people around the world, major outbreaks of influenza in 1957-1958 and 1968-1969 killed more than a million people each. The most virulent strains hit about every 30 years[32].* We see that these stronger and thus better strains of the virus die off within a limited span of time i.e. within a few months or a year. This would be quite contrary to the theory of evolution by natural selection as we see a stronger and thus better strain of the virus not able to survive whereas its milder and thus weaker form survive and spread among the host population.

"But when we bear in mind that almost every species, even in its metropolis, would increase immensely in numbers, were it not for other competing species; that nearly all either prey on or serve as prey for others; in short, that each organic being is either directly or indirectly related in the most important manner to other organic beings"[33]; We can say that this is a form of conservation where both the prey and the predator have enough features to keep each other in control in limited numbers, and neither have features to completely destroy the other i.e. for the predator to completely destroy its prey or the prey being so developed that it can always escape its predator. This is very important

feature for both the prey and the predator for their survival and evolution, for in all other cases both the predator and prey would not be able to survive i.e. if the prey always escaped its predator, the predator would not survive due to lack of food; and also it would lead to the uncontrolled growth of the prey leading to the destruction of the prey's environment and food source and thus the prey also will not be able to survive; if the predator always caught its prey, it would destroy all of its prey and thereby destroy its food source and the predator would not be able to survive. This is for two species, but would be equally applicable for complex interactions between many species, as it exists in our environment.

As per US National Library of Medicine and National Institutes of Health, annually in US, there are over one billion cases of common cold – a mild infection, tens of millions of cases of influenza resulting in 114,000 hospitalizations & 36,000 deaths, about 17,500 cases of bacterial meningitis – a deadly disease requiring immediate hospitalization[34]. We see that common cold, a mild infection which does not destroy its environment – the humans, is widely distributed as it conserves its environment; Influenza, a stronger infection, infects a much lesser population i.e. has a feature that limits its spread among the host population, thereby after weakening or destroying a part of its host population, ensures that the rest of the population is intact and that its host is intact and hence ensures its own survival. Whenever the virus became deadly, like *Spanish flu in 1918-1919, major outbreaks in 1957-1958 & 1968-1969*[32], the organism could not conserve its environment and died off within a limited span of time. Similarly, meningitis, a deadly infection, sustains itself in the environment by ensuring that its spread is very limited. Thus it is very important for the survivability of any organism to conserve its environment, or the damage they do to their environment should be within the limits that their environment can sustain itself. So all types of organisms – mild, strong, deadly etc., can survive, provided they have features to conserve their environment. Whenever an organism evolved so that it consumed or damaged its environment more than the environment could withstand, like the Spanish flu organism, the organism itself was eliminated. We can compare Spanish flu organism with the meningitis organism. With about 40 million deaths due to Spanish flu in 1918-1919 and about 18 thousand annual meningitis infections in USA, we can infer that Spanish flu was highly infectious whereas capability of meningitis to spread is very limited. Both are very deadly, probably meningitis deadlier, but whereas meningitis can sustain itself in the environment, Spanish flu could not survive. The difference, probably, being in their features to conserve their environment. Meningitis, with a very limited spreading capability, ensures that its environment remains intact, whereas the Spanish flu, by being both infectious and deadly, damages its environment, and ultimately the organism cannot sustain itself.

Sr. No.	Infection	Annual infection in population	Population	Annual Infection ratio in population (%)	Case-fatality ratio (%)
1.	Spanish flu[#]	extinct		20%	22%
2.	Plague[*]	2861	29,546,766	0.0001 %	70 %
3.	Yellow fever	61	28,674,757	0.0002%	50.8%
4.	Tetanus[*]	0.4	100,000	0.0004%	30%
5.	Japanese encephalitis[*]	40000	1000000	0.002 %	30 %
6.	Typhoid[*]	150	100,000	0.15 %	7.5 %
7.	Hepatitis B[*]	260,000	5,000,000,000	0.5 2 %	10 %
8.	Diphtheria[*]	20,000,000	5,000,000,000	0.4%	2.75%
9.	Lassa virus[*]	200,000	5000	0.2 %	2.5 %
10.	Influenza[+]	39,000,000	301,139,947	13.3%	0.09 %
11.	Malaria	10,712,526	39,384,223	27.2 %	0.13 %
12.	Common cold[+]	1000,000,000	250,000,000	400%	~ 0 %
13.	Useful micro-organisms in human digestive tract and human skin	Present all the time	Present in the entire population all the time	5200% (assuming organism cycle of 1 week)	0 %

Sources: + US National Library of Medicine and National Institutes of Health * CDC centers for disease control and prevention #World Health Organisation, Bull Hist Med.

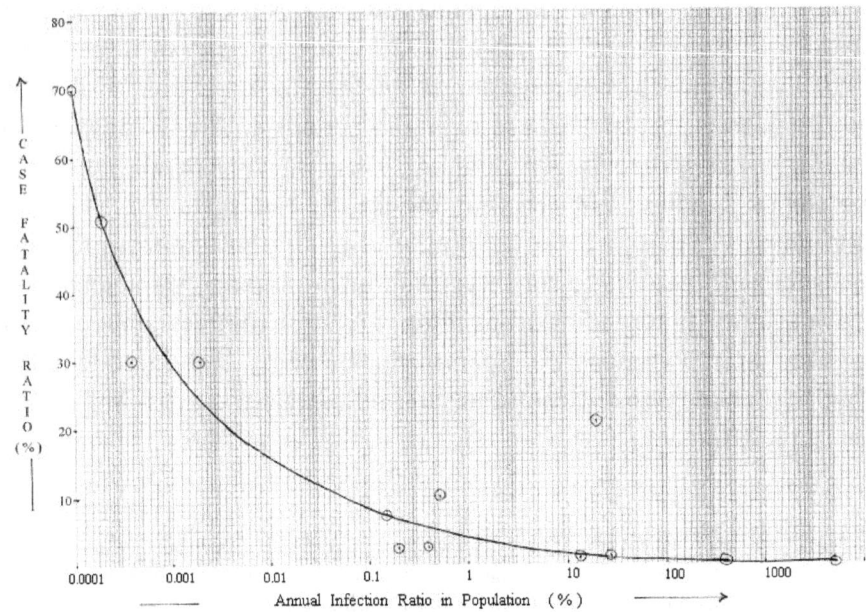

Annual Infection Ratio in Population (%)

In the above graph we see that the Spanish flu virus, with case-fatality ratio of 20% & annual infection ratio in population of 20% is outside the pattern of existing infections, and is not able to survive. In snakes, there are nearly 3000 species of snakes distributed throughout

Sr. No.	Type	Population density	capability
1.	Brahminy blind snake	very high	Very mild
2.	Northern rough green snake	430 / ha	mild
3.	Rough earth snake	348/ha	mild
4.	Southern Ringneck snake	900/ha	mild
5.	Eastern worm snake	100/ha	mild
6.	Corn snake	less	deadly
7.	Black rat snake	0.23/ha	deadly
8.	Southern copperhead	7.5 / ha	deadly
9.	cobra	less	deadly
10.	Python	less	deadly
11.	Timber rattlesnake	0.08 / ha	Very deadly
12.	King cobra	Very less	Very deadly

Source: Patuxent Wildlife Research Center

the world. Only 500 of these species are venomous i.e. 16.67 %. Also deadlier the snake, lesser is its population density.

As per Darwin's theory, *Organisms with traits that give them an advantage over their competitors are more likely to pass on their traits to the next generation than those with traits that do not confer an advantage*. However, what we see above is quite inverse; as we see fitness level increase – in case of infections, the case-fatality ratio & in case of snakes, the strength of venom - their population density decreases. In case of Spanish flu, an infection with a high fitness level, which has advantageous traits over its competitors; it could not survive and became extinct.

Like the Spanish flu and meningitis organisms, we humans, with scientific advancement, have become all powerful and deadly for our environment - exploiting, consuming and destroying almost whatever parts of our environment we choose. So we must be very careful in our actions and ensure that we exploit / damage our environment only to the extent that the environment can withstand.

Now if the common cold organism were to change and evolve with a more useful feature; i.e. it becomes more deadly for its host; then it will be able to survive over a long period of time only if it also develops a feature so that its spread become limited, thereby becoming less infectious; or on acquiring a useful feature, it ensures that some of its other features become less or acquires a handicap, so it remains mild and does not acquire a deadly form. Thus it is important for any evolving organism to have feature to conserve its environment, or feature of conservation is an important evolutionary force.

Let us consider a plant growing in a place which receives limited amount of water. If the plant evolved which required large amounts of water, it would dry up its place, and would not be able to survive. So only those varieties of plants will survive and flourish in that area, which can sustain themselves with limited amounts of water.

Although we see evolution or changes in organisms taking place in nature and in laboratories by changing environmental conditions; the species we see in our environment today are those which have stood the test of time and which can stay in equilibrium with others in their environment i.e. which do not damage their food resources or environment in which they grow, directly or indirectly, over a long period of time.

While this approach is quite simple, it can have big implications in understanding of the evolution process – of how species evolved to their present state, in present day understanding of evolution of new diseases, extinction of many species, and how our present day actions will influence us and our environment in the future.

Chapter 4
Conservation:
An Evolutionary Force?

We can say evolutionary force is a mechanism of nature by which different species evolved. Changes in species can be due to:

Mutation - new allele arises by physical change in structure of DNA (copying errors during meiosis)

Genetic drift - random change in allele frequency by chance, important mainly in small populations (variance as sample size)

Isolation - two populations that exchange members will not tend to diverge genetically. Isolated populations can diverge due to drift or natural selection.

Adaptation – when environment conditions of a species change

Migration - when a species migrates or is taken to a new environment

Above mentioned methods of change have been studied and analyzed, and natural selection has been shown to be the evolutionary force by which individuals in a species are selected for survival and propagation of their species. However, another important evolutionary force could be conservation or capability to live with others in their environment in equilibrium, without damaging their food source or the environment in which they grow, directly or indirectly, over a long period of time. We can say that, organisms have traits, got by selection, which make them fit to survive, and at the same time they also have traits, got by conservation, which make their environment capable to survive. Hence we can say that, conservation force can be quite different from as we view natural selection, as the traits of an organism existing due to conservation, directly ensure, not the survival of the organism, but the survival of its environment. Some features of conservation force can be:

* Traits got by selection directly make an organism fit to survive, whereas traits got by conservation indirectly make an organism fit to survive, as they ensure that their environment remains capable of surviving and thus ensures their own survival. Although the traits are of the organism, they ensure not that the organism is capable of surviving but that its environment survives.

* Like selection force, conservation force acts upon a set of traits of the organism.

* We may see organisms, survive and grow for many generations, which are otherwise not fit to survive from conservation point of view i.e. they have features that will not allow their environment to survive. Initially its population size will be small and its effect will only be felt when it has substantially destroyed its environment. This period of survival may be a few months or a year for organisms like Spanish flu, but may be hundreds or thousands of years for those with longer life cycles, slower growth or bigger environment.

* A trait of an organism may be acted upon by both selection and conservation forces. Whereas selective force will ensure that a trait is developed to a particular level so that the organism is fit to survive, the same trait may be acted upon by conservative force so that it does not develop to such a level that it begins to damage its environment. As an example, let us take the reproductive feature. Whereas a minimal reproductive capability is required for the survival of the species, too much reproductive capability can damage the environment of the species and so in a way conservation force can limit the reproductive

capability. So selection and conservation forces can act on the same trait in opposite directions and in some way counter act with each other.

I have preferred to use conservation force to be different from selection force as:

* Selection force is directly responsible for the capability of an organism to survive whereas conservation force is responsible for its environment to survive and thus indirectly responsible for the organism to survive.

* We sometimes see organisms like Spanish flu evolve and then fade away. What can we call this evolution of the organism due to, and again, what force is it due to which the same organism with the same traits fades away? We can say organism evolved due to selection but faded away because it could not conserve its environment.

* We can generally think of selection as a relatively quick process (it starts within the life period of the organism); whereas conservation can be the effect of the traits of an organism on its environment over a long period of time sometimes taking effect after several generations of the organism. Hence I feel it is better to make a distinction between selective force and conservative force.

I had taken examples where we saw clear interaction between predator and prey, and different types of predators acting on a single prey i.e. humans. However, this approach may be applicable to all species in general, but the analysis will be more complex due to networked and complex interactions between predator and prey. The term organism, used, also implies its group, species etc., unless otherwise stated.

Chapter 5
Equilibrium level

Fitness level: It can be the ability to survive in its environment and can be measured by:
- survival or mortality selection
- family size or fecundity selection – production of mature offsprings.

Conservation level: It can be the ability to let its environment survive and grow, and can be measured by:
- Fitness level – greater the fitness level, more will be its environment consumption / damage and population. Thus greater the fitness level, lesser will be its conservation level.
- Usefulness of its products / contribution of the species to the growth of its environment – greater the usefulness, higher is the level.
- Efficiency of its traits – for the same fitness level, greater the efficiency, lower will be the energy and food requirement, and hence lower will be the damage to its environment. Greater the efficiency, higher will be the conservation level.

Level of an organism: the total level of an organism would depend upon not only on its ability to survive but also its environment to survive. It can be measured by:
- fitness level
- conservation level

In a stable environment, an appropriate fitness level is maintained. The way to evolve is by increasing its role in growth of its environment & increase in efficiency of its features. Increasing efficiency leads to perfection in its various features. **This could be the**

reason why we see so much perfection in nature, with each species having developed features to perfection suited for their role in their environment.

Band of survival: Species survive in a narrow band of fitness level, whose lower limit or level of fitness is determined by its ability to survive in its environment and its upper limit by its capability to let its environment survive. The individuals of a species have a diversity of traits with a total level within this range of equilibrium. The range and distribution of fitness levels – surviving capability and population level, within the species will be such that the species will both be fit to survive and let its environment survive. In the process of evolution and struggle for survival, the capability to conserve or retain one's resources for survival is as important as the capability to exploit the resources and defend oneself. Rather than fit or fittest, species have to be ecologically fit.

Cow – eats grass, green plants, leaves etc., and produces dung which is useful for the growth of grass, plants etc.

Cow

www.love4cow.com

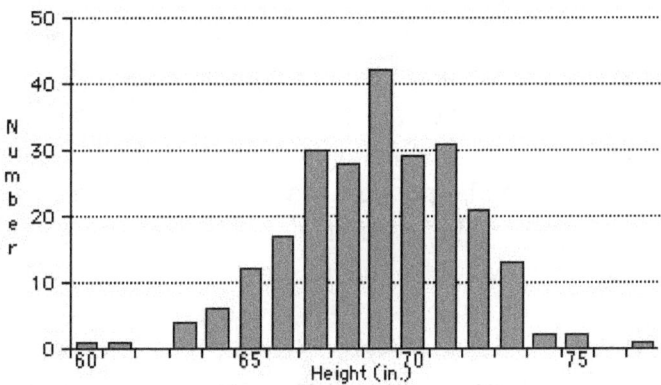

Height distribution of male secondary-school seniors[35]

If we see the variation of traits, most traits, e.g.,

- height
- body weight

vary in a continuous way from one extreme to another. A plot of the distribution of the trait in a population often produces a bell shaped curve like the one above, that shows the distribution of heights among a group of male secondary- school seniors[35].

We see this variation of traits remains stable over long periods of time, which would not be true if natural selection were to act, in which case we would expect the graph or trait distribution to shift in the direction of more fitness. However, we do see changes in trait distribution when there are changes in equilibrium level due to changes in environment conditions.

The lower limit for the fitness level of individuals in a species of particular population size would be the limit at which the organism will be able to survive, and the upper limit will be the limit at which its environment would be able to survive. Probably the organism in about the mid-range level would be best suited for its environment by being ecologically fit and being both capable to survive and letting its environment survive. The species as a whole would be ecologically fit.

Also stronger organisms may exist but under different population size (lower population or growth rate so that its fitness level remains same) and thus have different features and may exist as a different group. Thus there is tradeoff between higher consumption by an organism leading to decrease in its population or lower growth and benefits of its higher strength. It may be possible that a weaker organism i.e. species as a whole, may be able to survive in the environment whereas a stronger organism may not be able to survive.

Everything tries to achieve equilibrium in any ecosystem or sub ecosystem, and a change in any of the components of the ecosystem disturbs the equilibrium and changes take place in the ecosystem, to achieve a new state of equilibrium. From the purpose of surviving and growing in an environment, an organism living in an ecosystem or sub ecosystem is equal to all other organisms sharing its ecosystem. Some organisms may be more complex and more developed than others, but they are all equal from the point of view of capability to survive and grow in their environment. Here I would like to pose a question – why are there no bacteria or viruses which would destroy the whole world - the reason can be that these organisms live in a state of equilibrium with their surroundings, and the survival of others is necessary for their own survival.

Every organism lives in its own eco-system, the size and scope of which vary from species to species. The ecosystem maintains an equilibrium and all organisms must have a fitness level within the range of equilibrium so that they remain fit to survive and let their environment survive. The range and distribution of fitness levels within the equilibrium level will contribute to diversity among its species.

Some examples here have been taken for the purpose of illustration only and data is not actual as it might be difficult or not possible to measure. Let us consider the example of monkey. If say when interacting with a lion, with its senses, it can identify a lion and it needs a running speed of 20 km/hr to have enough time to run to a safe place before the lion can reach it; Then, both, a monkey with a running speed of 28 km/hr and a monkey with a running speed of 22 km/hr will be able to survive, unless there is change in equilibrium level like if the lion can run faster. Thus, as per the example, a minimum running speed of 20 km/hr would be necessary for the monkey to

Monkey

Photo: Neeraj Mishra © www.indianwildlifeimages.com

survive, and all those having greater running speeds would contribute to diversity among the monkeys. **This diversity among organisms is crucial for their survival**, as it only shows the capability of the species to survive in more adverse conditions like when the weather is more adverse or when one of the species in its environment becomes superior, etc.. Taking the above example of monkey, if the running speed of lion became faster and the minimum speed the monkeys should have to survive becomes 23 km/hr, then the species of monkeys will only be able to survive if there is diversity among the monkeys and there are many monkeys in its species which have a speed of 23 km/hr or more.

Equilibrium level: Every organism has qualities, features and abilities, the net sum of which must be equal or greater than the minimum fitness level for it to survive and grow but lesser than the maximum fitness level for its environment to survive. The individuals within this range of fitness levels will contribute to diversity within the species. The equilibrium level will thus comprise of the range & distribution of fitness levels – surviving capability and population level, so that the species is fit to survive and also lets it environment survive. Defining and giving a value to this level will not be simple due to numerous factors like various species, environment conditions like temperature, soil, water, etc., and may vary for different species. The equilibrium level of environment will include level of various species. One species is more developed or complex as compared to another, but I would not say that one species is superior to another. What I suggest is that, there is no perfect organism and every organism within an ecosystem is equal to others in its ecosystem from the view of survival and growth in the ecosystem. From the point of view of environment, what is more important is the total effect of all its features and its ability to survive and grow in the environment. When the environment conditions become adverse due to harsh weather (drought, extreme temperature etc.), increase in population, emergence / introduction of a better species, etc., the equilibrium level will rise. When the environment conditions becomes more hospitable like abundance of food supply, favorable weather conditions, abundant water etc., the equilibrium level will decrease.

So all organisms of a species in the equilibrium level, live in equilibrium within its eco-system, until a change in environment occurs. This causes a change in equilibrium level, thus bringing many previously fit to be unfit or previously unfit to be fit, depending on whether the equilibrium level has been raised or lowered. The environment change can be natural like weather, flood, drought, etc., or change in one or more species in the environment. With increase in the equilibrium level, the total of the various features of every organism, in the environment, must be in the new equilibrium level. This can lead to development of new features in species. An increase in equilibrium level will lead to increase in level or capability of existing features in some organisms; in others, new useful features, obtained by mutation, which were of not much importance till now, would become essential for the survival of the organism. In those that survive, the total sum of the features of the organism will be in the new equilibrium level.

The changes in environment can fine tune the development of a new and useful feature; an increase in equilibrium level may bring out a new feature or increased level of existing features; next a decrease in equilibrium level will bring about diversity in the new feature, and again an increase in the equilibrium level would bring about development of the new feature, thus fine tuning the development of the new feature. Further increase in equilibrium level will lead to more development of the organism, although there would be lesser diversity within every species. Thus there will be continuous accumulation of new features, or change in level of existing features, among the various organisms in the environment. This process of continuous accumulation of new features would ultimately lead to emergence of more complex and developed species.

Even without change in equilibrium level, when new features are added by mutation, and if they are useful to the organism, even though contributing very little to the overall fitness of the organism; it would cause a larger number of organisms with the new feature to be above the equilibrium level. As with the new feature, many with lesser developed other features (which without the new feature were not fit to survive), will now have the net sum of their features in the equilibrium level and thus fit to survive. Thus more individuals in the species with the new feature will survive and we will see a larger number of individuals in the species with the new useful features. Also a new feature can help an individual to go to a new environment, and if the level of the individual in the new environment is in the equilibrium level of the new environment, then the individual will survive and grow there. We will see the emergence of a new species when the individuals with the new feature begin to form their own separate group and grow & flourish within that group.

So we will see the emergence of new species – one with a new useful feature but lesser developed normal features, another with normal features and even one with an unfavorable feature but more developed normal features. However, the ones with new useful feature will be greater in numbers - having individuals with normal features plus individuals having level of features between normal and the required lesser developed features; and also have more diversity in their features - having normal diversity in the features plus those level of features between normal and required lesser developed features. Ones with new unfavorable feature will be lesser in numbers, having individuals with normal features minus individuals having features between normal and the required more developed features; and also lesser diversity in their features, having normal diversity in features minus those level of features between normal and required higher developed features. **This will lead to widespread propagation of new useful features and limiting unfavorable features.** With new useful feature, the average level of other features in the new species would be lesser so that the capability of species remains within limits and is ecologically fit. In both cases i.e. incorporation of both useful and unfavorable features results in changes in other features in the organisms, so that organism remains ecologically fit. E.g. a strain of influenza virus may give rise to many new strains, which are equally fit and each bring out different features to the environment. Each of these new strains may further give rise to other newer organisms leading to evolution of new species but which are equally fit for their environment & bring diversity to the environment.

To maintain similar capability, as organisms become bigger and complex, they have to be more developed. Comparison can be differences between a high rise building and single storey building – to withstand similar environmental conditions, design of high rise building has to be more precise, better design, workmanship more accurate, superior material have to be used as compared to construction of single storey building. Similarly, as organisms become bigger, to maintain similar fitness level, they have to be superior in design, material, construction, as compared to the simpler and smaller organisms.

When an organism develops a new useful feature, it can lead to the emergence of a new organism with a new useful feature but maybe some other feature lesser. This could be the reason why we see so much diversity in our environment, with each organism having its own special feature to survive - being better in some features and lesser in others, so that all organisms remain relatively equal in their ability to survive in their environment.

When the environment conditions deteriorate and equilibrium level increases, we will have some individuals with the new useful features while others with the normal features more developed. This will lead to the emerging of two different organisms, one with the new useful feature and the other with normal features more developed. However if the new feature is so useful to the organism, that even better developed normal features of organism cannot match the effect of the new feature or cannot bring the organism to the new equilibrium level, then we will just have the new feature incorporated into the species, without any other organism emerging.

Equilibrium is also maintained within species. A species can form various groups and sub groups among themselves and stay in equilibrium. They can have different habits, behavior, food sources, etc. the sum total of which will also contribute to their level or capabilities, and changes in equilibrium will see some groups better suited than others. Thus those habits, behavior, food sources, etc. of the organism which can keep it in the equilibrium level of the environment will be incorporated in the organisms, while others will fade away. Also the same organism can have different habits, behavior, food sources, etc. in different environments to be suitable for the different equilibrium levels of those environments.

Thus we can say that there is no perfect organism and every organism within an eco-system is equal to others in its eco-system from the viewpoint of survival and growth. There might be some which might be weak and thus unfit to survive in a particular ecosystem but strong and fit to survive and grow in another ecosystem. There might be others which might be too strong for their ecosystem, leading to their strong growth which at the same time destroys their ecosystem and ultimately their source of shelter, food and energy, and thus in the end leading to their own destruction and elimination. Therefore in the end we can see common only those species and life forms in the ecosystem which can grow and stay in equilibrium with others in their ecosystem.

As development of features is limited by conservation level, we will find development of features in direction so as to be able to exploit resources which have remained unexploited. The features will develop to the level where the resource will not be damaged or destroyed. This results in the various species, each having special features to exploit one or more of the various resources in the environment.

As conservation level becomes a limiting factor, hence rather than increased capability, evolution would lead to improved efficiency i.e. using same energy or food intake to have better performance or with similar performance to use lower energy or lesser food intake. The evolution in humans has led to walking upright, increased intelligence rather than increased height, strength etc..

As evolution takes place while maintaining the fitness level in limits, could be the reason that species remain stable over long periods of time and evolution takes place at a slow pace.

Chapter 6
Ecological selection vs Natural selection

The major difference with this approach to the development and emergence of organisms as compared to theory of evolution by natural selection is that: In natural selection, species are believed to have evolved by continuous selection from groups of organisms, the better of which survive, whereas in ecological selection, species evolve by continuous selection by being better both for itself and its environment; this approach of equality or equilibrium in environment is based on the theory that all organisms presently living are equal with regard to their capability to survive and grow in their environment, and maintain a state of relative equilibrium in their environment; and emergence of complex and developed organisms take place by new emerging organisms again achieving equilibrium in their environment or any new environment they may move to, and maintain a state of equilibrium with all other organisms in their environment. Now all organisms having a favorable feature or features for their ecological niche, besides inhabiting their niche in the environment also share other parts of the environment with other organisms. So if we consider the organism as a whole and include the effect of its favorable traits in its ecological niche along with its capability in other parts of the environment besides its niche, which it shares with other organisms; we will find that like other organisms, it is also equally fit to survive and grow in its environment.

Taking Zahavi's handicap principle where individuals having unfavorable traits, are favored as mates and thus favored to pass on their genes. The investment that an animal makes in signals is similar

49

to the 'handicaps' imposed on the stronger contestants in a game or sporting event. For example, the removal of the superior player's queen in a chess match, extra weight the swifter horse must carry, or the score of several strokes that the more accomplished golfer starts with.

Peacock

'Most people have seen and admired a peacock, spreading and quivering his enormous tail – a fan of glistering feathers, adorned with blue and green "eyes". But to be able to put such shows, peacocks have to drag massive tails around most of the year. By managing to find food and avoid predators, despite such a burden, a peacock proves that he is the high quality mate that the peahen is seeking to father her future chicks.

Like the artic fox with red bull's eye, an unfavorable trait for the artic fox; if these artic foxes with red bull's eye chose to form their own group and select within that group, we would see the emergence of a new group which has an unfavorable trait but still equally fit to survive and grow in its environment. This is true for all species and we

Arctic Fox

frequently come across unfavorable traits in almost all species. 'Why we take risks – What good does a huge tail do for a peacock? Peacocks carry splendid but cumbersome tails twice the length of their bodies. A bigger tail requires the male to waste huge amounts of energy, inhibits his ability to fly and makes him more vulnerable to predators. Why would Irish elk have developed antlers with a 12-foot span?'[36].

So the question is – why were those traits which are unfavorable not lost during course of time, as we would expect them, according to theory of evolution by natural selection. The reason can be – species develop and incorporate unfavorable traits also, as long as the net result of all their features is enough for them to survive and grow in their environment. The handicaps or unfavorable traits might be necessary for the conservation and survival of their environment.

The ecological model presented here is a model of evolution where the fitness level is bound within limits. A model of evolution based on perpetual growth and development would be a model where organisms ultimately destroy the very place they inhabit. Sustained growth in fitness is unsustainable. A model of evolution based on natural selection or survival of the fittest would be a similar model of perpetual growth and development and which would ultimately lead to the destruction of the very place they grow. However, this is not so and we see relatively stable ecology over long periods of time.

So rather than evolution based on continuous improvement in capabilities of species, the described model of evolution has capabilities of organisms bound by limits – the lower limit enough for it to survive in the environment and higher limit for ensuring it does not destroy the environment it depends on. An improvement in a feature in an organism / species would be accompanied by one or more of its other features becoming less, so that its net capability is within the limits. It explains why there is diversity and why and how such diverse and complex organisms evolved.

Natural selection deals with survivability whereas ecological selection, where fitness level remains within limits, deals with both survivability and sustainability.

Every organism has a life span where it grows to its maturity to its fullest strength and capability and then starts to weaken till it dies. Organisms having better genes could be selected from a group of organisms, if they were of similar development i.e. if they were of the same age (or in similar time frame as compared to their life span i.e. all organisms are fully matured, or all organisms are of 50% maturity size or all organisms are of 60% over the maturity age). However in reality, when there is a competition between species, the young and thus the not fully developed or the aged and thus the weak are the ones to be normally eliminated. *According to Minnesota Department of Natural Resources, wolves end up surviving on the most vulnerable individuals in the deer population, such as the very young, old, sick or nutritionally compromised, because those are the ones they can catch*[22].

Deer

Photo: Neeraj Mishra © www.indianwildlifeimages.com

Besides organisms too weak, organisms too strong for the environment are also not common. Take the example of organisms causing plague. These organisms kill almost all the individuals they infect, thus destroying their source of nourishment and shelter. Thus in course of time, either they will destroy themselves and become extinct, or will have features to infect only a few in a group or stay dormant and infect after long periods of time, so that the environment and the host they infect return to their original state. Hence, over a period of time, we will find only those organisms common in the environment which can stay in equilibrium with others in their environment, although many might have emerged and eventually become extinct in spite of being stronger and thus better fit, than others in their environment.

When an organism is taken from a region and introduced to a new region or environment, then either that organism is unfit to survive in the new condition, or is so deadly that it has a very rapid growth and destroys many of the original species in the new region. This can be explained as, in the original environment, the organism was in equilibrium with other species, but when moved to a new region, chances that the new region has exactly the same equilibrium level as the old region are very less, so either it is unfit or becomes deadly in the new environment.

Development of new features

When due to mutation, a useful feature is incorporated, which occurs once only in many mutations, initially the feature is very basic and not developed. With this new undeveloped and very basic feature, for the organism to stand out from other members of its species and survive as compared to others in the species and propagate its new feature, it will be difficult to comprehend. Another way of looking at it is, when a new and useful feature is incorporated, though very basic and undeveloped, with the new feature it will remain at the equilibrium level even if some of its other features are less developed. In any environment there will be many in every species below equilibrium level or become below equilibrium level when environment becomes adverse. Now any organism which gets a new useful feature, even though minutely small, and if it is below the equilibrium level, with the new feature, it might have the sum of its various features equal or above the equilibrium level. Thus an organism which was below the equilibrium level, and thus unfit to survive, would due to a minutely developed useful feature, might no more have a level below equilibrium level and thus be fit to survive and grow. Also a new feature might help it to move to a new environment and survive and grow there, if it is in the equilibrium level of the new environment.

So we will see a larger proportion of the population with the new feature survive, as compared with the proportion of the population without the new useful feature surviving. When it reproduces and this feature is passed to the next generation, similarly more of the young ones will survive, as even those with lesser other features, (which without the new feature would be below the equilibrium level and unfit to survive), would now be in the equilibrium level. Thus more of the organisms with the new feature will survive and the percentage in the species with the new and useful feature will increase. When individuals with the new feature begin to form their own separate group and grow & flourish within that group, we will see the emergence of a different group or species.

When there is a change in the equilibrium level of the environment and the equilibrium level becomes higher, some organisms whose other features are more developed will survive while others with the new feature will survive. So the new feature which was not an essential feature of the original organism becomes an essential feature for the organism to survive without which the organism will be handicapped.

With further increase in equilibrium level, some in the species will have the new feature more developed while others will have its other features more developed, and we see the emergence of two types of individual in the species. However if the new feature is so essential that the beneficial effect of the new feature cannot be compensated even with more development of other features, then just the new feature will be incorporated in the species. Thus there will be a slow and gradual accumulation of new and useful features in various species, whether in the same species or in the new species emerging.

Well, we can ask, why we don't see different level of development of the new feature in the different organisms. With increase in equilibrium level, the features of an organism become more developed, and all those with lesser developed features, have their level less than the equilibrium level and thus become unfit to survive. Thus we will see the feature developed to a level which is necessary to keep the organism at the minimum fitness level, and above which there is diversity among the organisms for that feature.

This may also explain why some species become extinct, as with increase in equilibrium level, if the organism with the most developed feature is also below the equilibrium level i.e. the development of the feature in the organism has reached its maximum and can develop no more, it becomes unfit to survive in the environment. Species with lot of diversity among its members have a much better chance of long term survival and those with less diversity will have a tendency to become extinct. Species with more diverse features are more likely to survive unfavorable conditions as at least some of its individuals will be fit enough to face the unfavorable condition. On the contrary, if diversity was not there, and all individuals were alike, species would become extinct as none would be able to face a changed and unfavorable condition. Thus some species become extinct while others pass the test of time. We can ask a question, why we don't see a creature similar to man, but say with no hearing and much advanced eyesight; the answer can be that even with maximum development of other features like eyesight, strength, intelligence, etc. but no hearing, will not bring it to the minimum required fitness level and will be unfit to survive in the environment. Many a times we see species evolve from one to another, with intermediate species not fit to survive and becoming extinct.

Diversity

When a new species emerges from another species, many times we see both the species survive and flourish; which would not occur if the new species were superior to the old species. The emerging species coexist in equilibrium with other species in the environment and are equal to others in their environment with respect to their ability to survive and grow in the environment. The different species use their own special abilities to exploit the varied resources and habitats of their environment. They bring about diversity to the environment.

Spurt in new Species

As per historical evidences, it is estimated that there was a sudden spurt in new life forms and species, occupying the various ecosystems of the environment millions of years ago. The sudden emergence of new forms of life into new ecosystems cannot be brought about by natural selection, with individuals competing for common resources. This can be better explained as organisms diversifying and occupying unoccupied systems. When organisms, by mutations, get features giving them capability to move to a new environment, they will survive and grow there, if the new environment is suitable. In the initial stages of emergence of life on earth, the number of species were very small; thus there were very few or no predators with lot of resources for nourishment, and thus easy to inhabit and survive. So initially, just getting a new feature, giving an organism capability to move to a new and uninhabited environment, was probably sufficient for it to inhabit the place. We could see a spurt in the emergence of a new species. Different life forms and species begin to emerge occupying their own part of the environment and living in equilibrium within their own ecosystem. The new life forms, to survive and stay in equilibrium, within their ecosystem, have features which are suitable for their ecosystem. The different species occupying the different parts of the environment have their characteristic features which are suited for the organism to stay in equilibrium with others in their environment.

There might be some which might be weak and thus unfit to survive. There might be others which might be too strong for their ecosystem, leading to their strong growth and destroying their source of shelter, food and nourishment, and leading to their own destruction and elimination. Therefore, in the end, we will see a wide diversity of life forms survive, which can grow and stay in equilibrium with others in their ecosystem.

Absence of intermediate forms

When a new form evolves, what happens to the original form? or why don't we find traces of the intermediary forms? A reason we can attribute to it is; that, till individuals in a group with a particular feature or a particular feature developed to a particular level, form their own group and select & pair within that group, they are a part of the original species contributing to the diversity of the species. They begin to emerge as a different form or species when they begin to form their own groups and grow within that group. We know that many times both the original species and new emerging species survive.

Example is Globorotalia crassaformis. There is a location in the South Pacific where this species gradually turns into a transitional species, G. tosaensis, and then into G.truncatulinoides. The gradual change took 500,000 years. We know that these are different species (and not just strange looking fossils of a single species) because both species still exist today[2].

Also when a species develops a feature which is so essential that increase in other features cannot compensate for incorporation of the new feature, the new feature is slowly incorporated into the species. The feature becomes more developed when there is an increase in equilibrium level of environment, either due to weather conditions becoming adverse, scarcity of food etc., or increase in level of a competing species.

When individuals of a species 'A' with a particular feature above a certain level, form their own different group and select & pair from that group only; it would lead to the emergence of a new species or sub-species 'B'. This would result in the equilibrium level

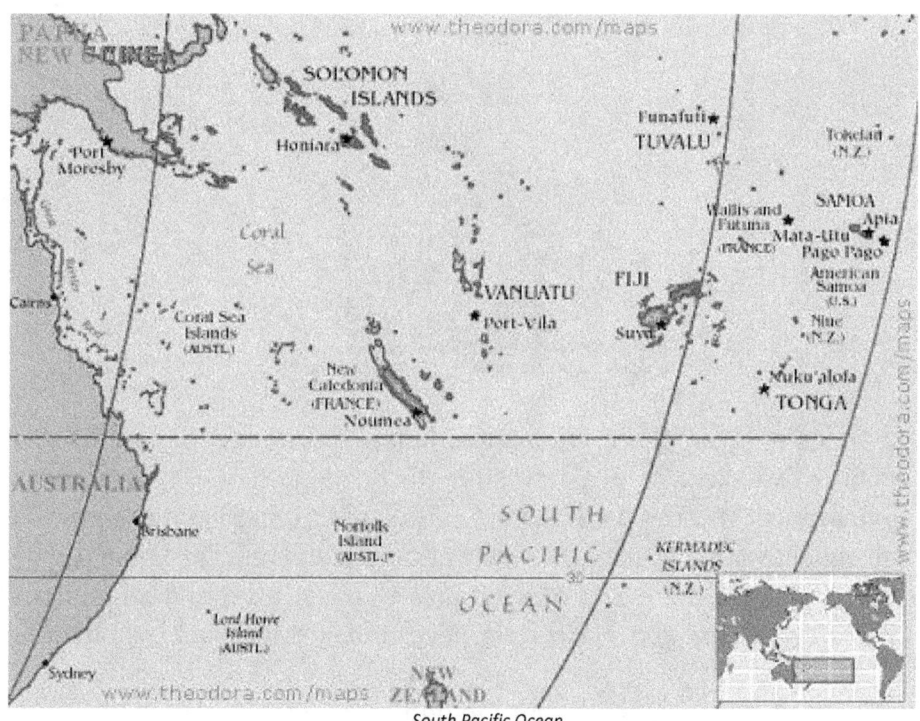

South Pacific Ocean
Map courtesy of www.theodora.com/maps, used with permission

of the environment remaining the same or increasing depending upon whether the level of individuals in the new species 'B' is same or higher than the level of individuals of old species 'A'. With the emergence of the new species 'B', there might be a change in equilibrium level, and the old species will survive if some or all of individuals in species 'A' are in the new equilibrium level, and its features will change or not depending on what changes there are in its environment due to the emergence of the new species 'B'. Again if some of the new species 'B' develop new features and form another group 'C' due to which there is change in equilibrium level and 'B' becomes unfit to survive whereas 'A' is fit to survive, then we will just see forms 'A' and 'C', with the form 'B' becoming extinct. Thus evolution may take place by not only increase in level of one species but also, at the same time, by bringing changes in species surrounding it, so the new species again remains in a balanced state with its surroundings.

Gorilla

Photo: Jessie Cohen © http://nationalzoo.si.edu

As an example, if we assume humans evolved from gorilla and man is weaker in physical strength to gorilla while greater in intelligence. This can be explained as, due to increase in intelligence, to be equal to other species, the body or muscular strength of men can be lower, so that there is a diversity in the body, form and strength of human which are above a certain level but much lower than that of

Homo habilis

www.geocities.com/palaeoanthropology/Hhabilis.html

The early discoveries of early hominid fossils were made at Olduvai Gorge, by the Leakeys. The fossils were thought to be slightly older than about 1.75 million years and in addition, the teeth were smaller and the brain was calculated to be significantly larger. Leakey believed that habilis was a direct human ancestor. Homo habilis means "handy man" and was suggested to them by Raymond Dart.

a gorilla. Also there is diversity in body and strength of gorilla but the minimum level to make it fit to survive is much greater than that of a human. Now we do not see a species between that of gorilla and man, although they might have existed in the past. Initially with the emergence of a new species 'Q' from gorilla, the new species would be greater in intelligence but lesser in strength. Again a new species 'R' may emerge from 'Q', which is more intelligent but less in strength than 'Q'. The new species 'R' will bring a change in equilibrium level due to which the species 'Q' may become below the equilibrium level and thus unfit to survive, whereas the gorilla due to its superior body and strength would remain in the equilibrium level. Similar changes may also have taken place in gorillas. This process might have continued to the present when we see just the gorilla and humans with the intermediate species becoming extinct.

Chapter 7
Ecological balance

The more the diversity of forms within a species, more are its chances of long term survival, as it increases the potential of the species to survive in more difficult and adverse environments. The potential of its various features to improve will be shown by the diversity of those features in the species.

Diversity in traits exists among the various individuals of a species. Change of environment brings about a new state of equilibrium, so many which are presently fit may become unfit. An example would be cigarette smoking. As statistics, we can say that a certain percentage of people would have throat or lung problems due to cigarette smoking. However, we will find some chain smokers smoking regularly say for over 50 years and be quite healthy, whereas some who have been smoking for only a few years get problems. This is due to diversity among people to the effect of cigarette smoke. This is true for all changes in environment like pollution, food, work habits etc. Organisms grow and survive in equilibrium along with others in their environment, and any changes in environment or way of living might be good or harmful, but what exists in the environment and our way of living will not be harmful as we have grown and developed around it. So we must be careful what changes we do to our environment.

We are never alone. After birth, we share our bodies with about a hundred trillion micro-organisms, even when we consider ourselves to be clean. The number is so large that micro-organisms outnumber the human cells of our bodies by about ten to one. Some of these microbes are mere fellow travelers, doing us neither good nor harm, but many are important to the way that our bodies work, and we need them because they can perform bio-chemical tricks that human cells have never evolved. The healthy gut contains between one and ten billion bacteria per gram of tissue. These tiny organisms perform several important tasks like digesting components of food that our own enzymes cannot attack. Recent research suggests that humans and their helpful bacteria have, over their long association, evolved ways of communicating with each other, so that the two very different types of organism can operate as a single, integrated system. One of the many ironies of human life is that it depends on death – the death of vast numbers of individual cells in an otherwise healthy body. In normal human, the cells that lie between the fingers kill themselves, causing the overlying skin to shrink back towards the palm and to take on the form of a glove rather than a mitten. Elective cell death is a common feature of permanent tissues, where it is used to eliminate excess cells. Many developing tissues over-produce cells initially and then use signals from other tissues to determine how many, and which, should be kept[24].

Communications and signaling between cells are important for proper growth of tissues and organs. Use of pesticides, insecticides, chemical pollutants, genetically modified foods may affect this signaling or health of beneficial bacteria on skin, gut etc.. *Thalidomide breaks down in the body to produce a molecule that inhibits the outgrowth of new and immature blood vessels. Sudden appearance of babies with deformed and shortened limbs was connected with thalidomide. The modular nature of a body constructed around individual organs builds itself using mainly local communications between its own cells. It refers to the rest of the body for only a few key decisions, such as when to begin building itself, when building should end*[24].

All organisms depend on each other for survival, each being food for others while at the same time consuming others as its food, starting from the plant species which convert carbon dioxide, minerals, water etc. to living matter, to the carnivores which feed on other animals and finally to the bacteria which convert the dead organisms into minerals, hydrocarbons, gases etc. to return them to the environment.

Any evolving species would ensure not only its own survival but also ensure that it does not destroy other species on which it depends.

Each species on which it depends further ensures they do not destroy the species on which they depend. Thus every species has a role in the survival of other species on which they depend, and as all species depend on each other directly or indirectly; survival of every species is maintained. *All species are locked in a life or death zero-sum game competing for the same, shared, limiting resources*[23].

There will be a balance between number of individuals in one species and number of individuals in another species, plant or animal, with which it interacts either by being its food source or it being its food. *Plants are less than 10% efficient in converting light energy to produce pant tissue. Ninety percent is lost as heat to the atmosphere. In the transfer of energy to the second trophic level, the herbivores follow suit, so essentially energy is degraded by 90% at each level from plants through herbivores to carnivores*[22]. Thus we can expect plants to be in much greater numbers than herbivores, which in turn would be in much greater numbers than carnivores. Every organism needs food and shelter for its growth and survival. However, in a specific ecosystem, there can be only a limited source of food supply and shelter and this limits the number of individuals of that species in the ecosystem. As the amount of food and shelter, in any environment, is sufficient for a limited number of individuals in a species; so till the number of individuals in a species is less than this number, there will be growth to reach this number. Any increase over this number will cause constraint over the food supply and shelter of the organism, thereby limiting their numbers. As all organisms interact with each other either directly or indirectly, there will be a balance between the number of individuals in each species in the environment, creating an ecological balance.

Chapter 8
Handicap principle

Amotz Zahavi

Zahavi's handicap principle – individuals in a species with unfavorable traits favored as mates - can be better explained with this approach. For individuals having unfavorable traits, in order to survive and propagate, they must meet the equilibrium level of the environment. Thus organisms with unfavorable traits, must be having other features more developed. The presence of an unfavorable trait and proof of survival becomes a proof that its normal features are more developed than the normal features of others in the species. These unfavorable traits are normally unfavorable for survival in the environment when interacting with other species and predators. But when interacting with members of their own species, these unfavorable traits do not show any unfavorable features but only show that its other features are more developed, and thus has a better position within the species and be a better and chosen mate.

For example, the artic fox with red bull's eye, to survive in the environment must have the total of all its features in the equilibrium level of the environment. The presence of the red bull's eye in the artic fox, an unfavorable trait for it when dealing with its predators, just shows that the red bull's eye artic fox has one or more of its features more developed than those of other members of its species.

Arctic Fox

With its more developed features and the presence of the red bull's eye, an unfavorable trait; but not an unfavorable trait when dealing with its own species, puts it in a more favorable and dominant position within its species and a more favorable mate. The red bull's eye thus becomes an identification mark for individuals with more developed features within the species. The artic fox has capability to develop further but only develops when accompanied by an unfavorable trait like the red bull's eye. The handicaps or unfavorable traits might be necessary, for the conservation and survival of its environment, and thus indirectly its own survival. The unfavorable traits limits the capability of the individual.

The investment that an animal makes in signals is similar to the 'handicaps' imposed on the stronger contestants in a game or sporting event. For example, the removal of the superior player's queen in a chess match, extra weight the swifter horse must carry, or the score of several strokes that the more accomplished golfer starts with, to make them at a level playing field with a lesser accomplished competitor. Whereas in games, handicaps are introduced by choice, in species handicaps become a part of their features.

'Black cats provide lucky break for researchers – New natural genetic resistance may be discovered – Dr. Stephen O'Brien, Eduardo Eizirik and colleagues were wondering what made cats black – not out of idle curiosity, but because such genes often confer protection against disease. Otherwise, animals with unusual coloring would go extinct.'[21] – An evidence that for survival what is important is not the increase or

Black Cat

decrease in the level of a single trait or feature but the total effect of all the features of the individual. If there is deficiency in one trait, like an unfavorable color, it is compensated by increase in level of other traits like increased disease protection.

Chapter 9
Our Earth and Universe

The conditions maintained on the surface of earth including temperature, water, atmospheric pressure & composition, is a delicate balance based on many feedback systems operating in the environment. Within limited variations in conditions, these feedback systems work well, but if changes are introduced which interrupt these feedback systems, its effects will not be easy to predict and whatever the change, it definitely will not be favorable, as we have developed and become adapted to this environment. Also, earth has enormous amount of energy in the form of carbon and high temperature below earth surface – if we were to exploit these energy for our use, what would be the resultant increase in earth's surface temperature is again difficult to predict. There is a balance between the solar radiation received on earth and amount of energy irradiated back from earth to space, and this balance maintains the prevailing surface temperatures on earth. Our earth's core has a temperature of more than 5000°C and a temperature of less than -60°C at a height of 10km above earth's surface. However, most of the hospitable regions of the earth's surface maintain a narrow range of temperature between little below zero °C to max of 55°C. Any introduction of extra energy in the form of burning of fossil fuels or releasing of energy from earth's core, or changing earth's surface features to reflect/absorb/radiate solar energy will have a direct effect on earth's surface temperature.

If we examine the history of our earth, we see that since the formation of our earth, its atmosphere has undergone many changes including change in composition from being highly rich in CO_2 to being rich in O_2. If we assume that there is negligible loss of atmosphere from earth to space, this CO_2 will all have converted into carbon and its compounds including hydrocarbons. The amount of carbon now in our earth will be enough to bring back the CO_2 in our atmosphere to a level of high CO_2 as existed earlier.

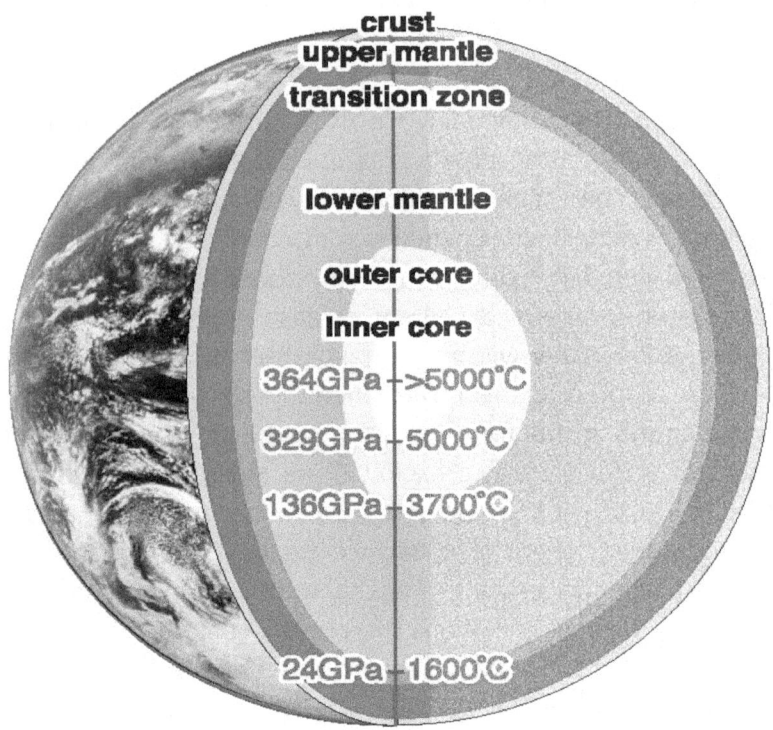

Stephen Hawking has said that *humans may lack the skill as a species to stay alive.* He had warned that the aggressive instincts of humans, coupled with fast growth in technology, may destroy us all by nuclear or biological war, adding that only a 'world government' may prevent this impending doom[38].

Human being is the only species which sets out to destroy the ecosystem without any balance. When we make a species extinct, what are we losing? – I believe it is unfathomable. Living organisms are composed of organic matter, cells, organs, etc., and made up of molecules which we can identify. But there could also be an aspect or characteristic we cannot detect or understand, an aspect which contributes to the functioning of the species. A dog can hear sounds we cannot hear, homing pigeons have impeccable navigation skills – till now we are not able to understand how. Species could function, communicate or detect in ways we cannot fathom because it's so different from our realm of understanding. Does the answer lie in dark matter, dark energy[*], or something else? Perhaps. We must accept that there are many things we do not know and behave accordingly. We must be careful in what we destroy or change, especially those we can never recreate.

We, as a species, do not understand or do not want to accept the scale of damage we doing to our environment and ultimately to ourselves. Dr. Jennifer Lavers of University of Tasmania recently did a study of Henderson Island, located more than 5000 kms from the nearest major population centre[37]. She found that it is littered with an estimated 37.7 million pieces of plastic. What's happened on Henderson Island shows there's no escaping plastic pollution even in the most distant parts of our oceans. Dr Lavers said most of the more than 300 million tonnes of plastic produced worldwide every year is not recycled, and as it's buoyant and durable, it has a long term impact on the ocean.

Thank you for reading my book. Comments will be appreciated and may be sent to pvallabh@gmail.com.

* Dark matter and Dark energy: *There is strong theoretical and observational evidence indicating that a mere 5 percent of the universe's heft comes from the constituent's found in familiar matter – protons and neutrons (electrons account for less than .05 percent of ordinary matter's mass) – while 25 percent comes from dark matter and 70 percent from dark energy. But there is still significant uncertainty regarding the detailed identity of all this dark stuff. So, if not protons and neutrons, what constitutes the dark matter? As of today, no one knows, but there is no shortage of proposals. That no* *one has yet detected a dark matter particle places significant constraints on any proposal. The reason is that dark matter is not only situated out in space; it is distributed throughout the universe and so is also wafting by us here on earth. According to many of the proposals, right now billions of dark matter particles are shooting through your body every second, so viable candidates are only those particles that can pass through bulky matter without leaving a significant trace[25].* It could be possible that as we see the world made of atoms and molecules, there exists another world made up of dark matter and dark energy.

References:

1. Handicap Principle, A missing piece of Darwin's puzzle by Amotz and Avishag Zahavi, Oxford Univ Press, 1997

2. First appearance of Globorotalia truncatulinoides: cladogenesis and immigration by Spencer-Cervato, C. and Thierstein, H.R., Marine Micropaleontology, v. 30, 1997 p. 267-291.

3. The Genetical theory of Natural Selection by R.A. Fisher (1930), Claredon Press, Oxford

4. Neutral Theory of Molecular evolution by Motoo Kimura – Cambridge Univ. Press 1985

5. Evolution, edited by Mark Ridley, Blackwell 1996

6. The evolution of sexual preference by R.A. Fisher (1915) pages 184:192

7. The structure of evolutionary theory by Stephen Jay Gould, The Kelknap press of Harvard University Press, 2002

8. Natural enemies, the population biology of predators, parasites and diseases, edited by Michael J. Crawley, Blackwell Scientific Publications 1992

9. Population cycles, the case for tropic interactions, edited by Alan Berryman, Oxford Univ. Press 2002

10. Predators and Predation, the struggle for life in the animal world, edited by Pierre Pfeffer, FactsOnFile 1989

11. Animal Conflict by Felicity A Huntingford and Angela K. Turner, Chapman and Hall,

12. Living Together, the biology of animal parasitism, by William Trager, Plenum Press

13. Minimum animal populations, by Hermann Remmert, Springer Verlag,

14. Life Unfolding, how the human body creates itself by Jamie A. Davies, Oxford Univ Press

15. The merits of Neutral Theory by David Alonso, Rampal S. Etienne and Alan J. McKane. TRENDS in ecology and evolution. Vol. 21 No. 8

16. Collapse, how societies choose to frail or survive, by Jared Diamond, Penguin Books

17. The Blind Watchmaker, by Richard Dawkins, Penguin Books

18. The selfish Gene, by Richard Dawkins, Oxford Univ Press

19. Guns, Germs and Steel by Jared Diamond, Vintage Books, London

20. Sapiens, A brief history of mankind, by Yuval Noah Harari, Harvill Secker, London

21. www.abc.net.au/news/2003-03-04/black-cats-provide-lucky-break-for-researchers/2690744

22. www.conervationnw.org/what-we-do/predators-and-prey/carnivores-predators-and-their-prey

23. The Unified Natural Theory of Biodiversity and Biogeography by Stephen P. Hubbell, Princeton Univ. Press 2001 Pg 6, 28, 316, 322

24. Life Unfolding by Jamie A. Davies, Oxford Univ. Press 2014 Pg 123, 144, 177

25. The Fabric of The Cosmos by Brian Greene, Vintage Books 2004 Pg 432, 433

26. J.G.Teer et al., in Transactions of the North American Wildlife and Natural Resources Conference 1991, vol. 56, pp 550-560

27. J.G.Kie, D.L.Drawe, G.Scott, Changes in Diet and Nutrition with increased Herd size in Texas White-Tailed Deer, Journal of Range Management 33, 28-34 (1980)

28. Biology Life on Earth by Audesirk, Audesirk & Byers, pub. Prentice Hall Pages 353, 313-314, 270-308

29. Biology Understanding life by Sandra Alters, pub. Mosby, Pages 488-508

30. Advanced Biology principles & applications by C J.Clegg with D G Mackean, pub. John Murray (publishers) Ltd., pages 652-678

31. Biological Science I & II by DJ Taylor, NPO Green and GW Stout 3rd edition pub. Cambridge Univ. press, page 879-904

32. World Health Organization, pandemic Influenza (http://www.who.int/mediacentre/factsheets/fs211/en/)

33. Darwin, C The Origin of Species, Edited by J.W.Burrow (Penguin Books, London, 1968) pp. 208-208

34. U.S. National Library of Medicine and National Institutes of Health, Medline Plus Trusted Health Information for You. (http://www.nlm.nih.gov/medlineplus/ency/article/000678.htm).

35. http://users.rcn.com/jkimball.ma.ultranet/BiologyPages/E/Evolution.html

36. DISCOVER, Vol. 22, No. 12 (Dec. 2001).

37. Jennifer L.Lavers and Alexander L.Bond, Exceptional and rapid accumulation of anthropogenic debris on one of the world's most remote and pristine islands, pub. PNAS 2017: 1619818114v1-201619818

38. www.deccanchronicle.com/science/science/040517/man-must-leave-earth-to-survive-says-stephen-hawking.html

www.ingramcontent.com/pod-product-compliance
Lightning Source LLC
Chambersburg PA
CBHW061444180526
45170CB00004B/1545